Report of Investigations 9690

A New Leak Test Method for Enclosed Cab Filtration Systems

John A. Organiscak and Michael Schmitz

DEPARTMENT OF HEALTH AND HUMAN SERVICES
Centers for Disease Control and Prevention
National Institute for Occupational Safety and Health
Office of Mine Safety and Health Research
Pittsburgh, PA • Spokane, WA

May 2012

This document is in the public domain and may be freely copied or reprinted.

Disclaimer

Mention of any company or product does not constitute endorsement by the National Institute for Occupational Safety and Health (NIOSH). In addition, citations to Web sites external to NIOSH do not constitute NIOSH endorsement of the sponsoring organizations or their programs or products. Furthermore, NIOSH is not responsible for the content of these Web sites. All Web addresses referenced in this document were accessible as of the publication date.

Ordering Information

To receive documents or other information about occupational safety and health topics, contact NIOSH at

>Telephone: **1–800–CDC–INFO** (1–800–232–4636)
>TTY: 1–888–232–6348
>e-mail: cdcinfo@cdc.gov
>
>or visit the NIOSH Web site at **www.cdc.gov/niosh**.

For a monthly update on news at NIOSH, subscribe to NIOSH *eNews* by visiting **www.cdc.gov/niosh/eNews**.

DHHS (NIOSH) Publication No. 2012–145

May 2012

SAFER • HEALTHIER • PEOPLE™

Contents

Abstract 1
Introduction 2
Examination of CO_2 Instruments 3
    Experimental Test Methods 4
    Instrument Test Results 6
Development of New Cab Leakage Sampling Methodology 10
    Experimental Test Methods 11
    Test Methodology Results 14
    Discussion of Cab Leakage Measurement Error 17
Conclusions 20
References 22
Appendix A. Leakage Model for Cab Filtration System 25
Appendix B. Initial Experimental Leakage Data From Cab Test Stand 28
Appendix C. Refined Experimental Leakage Data From Cab Test Stand 29
Appendix D. Propagation of Error Analysis for the Cab Leakage Testing
    Methodology 31

Figures

Figure 1. CAF cab test stand setup for comparative instrument testing 5
Figure 2. John Deere tractor cab setup for comparative instrument testing 6
Figure 3. Instrument measurements inside CAF cab test stand during test series 1 7
Figure 4. Instrument measurements inside John Deere tractor cab during test series 1 9
Figure 5. CAF cab test stand setup for leakage experiments 11
Figure 6. Leak testing the John Deere tractor cab 13
Figure 7. CO_2 concentrations for the first series of initial leakage tests 14
Figure 8. Initial leakage test measurements made on the CAF test stand 15
Figure 9. Typical CO_2 concentrations during refined leakage tests 15
Figure 10. Refined leakage test measurements made on CAF test stand 16
Figure 11. Cab leakage RSD_l estimates with a CO_2 instrument RSD of 0.05 19
Figure 12. Cab leakage RSD_l estimates with a CO_2 instrument RSD of 0.10 19
Figure A-1. Basic enclosed cab filtration system 25

Tables

Table 1. Carbon dioxide instrument specifications 4
Table 2. Initial instrument comparisons with reference gases 7
Table 3. Linear regression parameters of Telaire and Vaisala instruments as compared to the Sable instrument inside the CAF test stand. 8
Table 4. Instrument comparisons with reference gases after user recalibration. ... 8
Table 5. Linear regression parameters of Telaire and Vaisala instruments as compared to the Sable instrument inside the John Deere tractor cab. 10
Table 6. Average leakage and confidence intervals for CAF test stand experiments 16
Table 7. Average leakage and confidence intervals for John Deere 7820 tractor cab 17

ACRONYMS AND ABBREVIATIONS USED IN THIS REPORT

ASAE	American Society of Agricultural Engineers
ASTM	American Society for Testing and Materials
CAF	Clean Air Filter
CRADA	Cooperative Research and Development Agreement
CO_2	carbon dioxide
HVAC	heating, ventilation, and air conditioning
NIOSH	National Institute for Occupational Safety and Health
RSD	relative standard deviation

UNIT OF MEASURE ABBREVIATIONS USED IN THIS REPORT

ft	foot
ft^3	cubic foot
ft^3/min	cubic feet per minute
in	inch
L	liter
L/min	liter per min
m	meter
µm	micrometer
mL	milliliter
mL/min	milliliter per minute
min	minute
psig	pounds per square inch gauge pressure
ppm	parts per million
%	percent
sec	second

A NEW LEAK TEST METHOD FOR ENCLOSED CAB FILTRATION SYSTEMS

By John A. Organiscak[1] and Michael Schmitz[2]

Abstract

A new test method has been developed by the National Institute for Occupational Safety and Health (NIOSH) and Clean Air Filter (CAF) for quantifying the outside air leakage into environmental cab filtration systems. This method uses specially configured filter cartridges to remove carbon dioxide (CO_2) from the environmental cab's air filtration system. Real-time gas monitors are used to measure the outside and inside cab CO_2 concentrations after the cab reaches steady-state equilibrium conditions inside an unoccupied cab. Cab filtration system leakage can be mathematically determined using the measured cab penetration (inside to outside cab concentration ratio) and the special CO_2 filter cartridge efficiency.

Examination of several CO_2 sampling instruments for this type of testing showed them to be relatively precise, but exhibited noticeable variations in accuracy. These results indicate that frequent gas calibration checks of comparative sampling instruments would be needed to ensure their accuracy, which is not well-suited for cab field testing. Thus, a single-instrument, multiple-sample-location cab testing methodology was devised to eliminate multiple-instrument sampling biases and frequent calibrations during testing. This methodology was examined and can provide accurate measurements of filtration system leakage into enclosed cabs with a precise instrument at or near steady-state test conditions. The new leak test method provides cab manufacturers, cab service personnel, and industrial hygienists with a measurement tool to ensure environmental cab integrity and minimize worker exposure to outside airborne substances.

---

[1] Mining Engineer, Office of Mine Safety and Health Research, NIOSH
[2] President and Director of Research and Development, Clean Air Filter, Defiance, Iowa

Introduction

Enclosed cabs are an engineering control that can provide a safe, comfortable, and healthy work environment for equipment operators. Most modern day enclosed cabs have heating, ventilation, and air-conditioning (HVAC) systems for maintaining a comfortable temperature and a breathable quantity of air for its occupant(s). Various levels of filtration can be incorporated into the HVAC system to improve the ventilation quality of the air inside the cab by removing outside airborne pollutants such as dusts, chemical aerosols, and vapors. Outside air leakage around the intake filter into the HVAC system can notably diminish the cab's filtration system effectiveness [Heitbrink et al. 2003; NIOSH 2008]. A poorly sealed cab HVAC/filtration system can be difficult to recognize because of its concealed system components and the invisible nature of some airborne pollutants that can penetrate the cab.

Enclosed cab filtration system effectiveness has been previously studied and can be difficult to measure in practice. The American Society of Agricultural Engineers (ASAE) previously devised a consensus standard for testing a cab's particulate reduction factor and specifying a cab's performance criteria for pesticide applications. These procedures used optical particle counters inside and outside the cab to examine 2 to 4 µm ambient air particulate penetration into the cab as it drives along on an end-use tractor [ASAE 1997; Heitbrink et al. 1998]. An alternative test procedure that has been examined is particle counting inside and outside of a stationary vehicle cab parked inside a temporary enclosure filled with incense smoke contaminants [Moyer et al. 2005]. Other researchers have measured respirable dust mass concentrations inside and outside of mining equipment cabs during multiple production shifts [Organiscak et al. 2003; Cecala et al. 2003; Cecala et al. 2005]. Several of these studies have indicated that inconsistent and low particulate or dust concentrations can yield unreliable cab performance results between replicated cab tests [Heitbrink et al. 1998; Organiscak et al. 2003]. Also, internal cab particulate generation such as dirty floors, interior surfaces, and abraded blower motor brushes can also interfere with measuring external particle or dust penetration into the cab [Cecala et al. 2005; Heitbrink and Collingwood 2005].

In a response to develop alternative cab performance test methods, NIOSH entered into a Cooperative Research and Development Agreement (CRADA) with Clean Air Filter (CAF) of Defiance, Iowa to develop field test methods for evaluating environmental integrity of enclosed cabs. This research was conducted to develop an expedient, simple, quantitative, and reliable field test method for measuring air leakage into enclosed cab filtration systems. A field test was sought that used a measurable airborne agent around the test vehicle which would pose minimal health and safety risks to the user. Atmospheric gases have these desirable attributes and were considered for the leak test medium or tracer. Cab leakage testing research was ultimately conducted with carbon dioxide given that instrumentation and gas-absorbent media were readily available for filtration and methodology development. This research focused on conducting a timely stationary cab leak test for an unoccupied cab.

The initial development of this new leak testing concept is described in a previously published paper [Organiscak and Schmitz 2006] and several patents [Organiscak and Schmitz 2009; Organiscak and Schmitz 2010]. Enclosed cab carbon dioxide filtration and leakage experiments were conducted on a laboratory enclosure test stand to formulate and validate a testing methodology. During this research a mass balance mathematical model was developed and is shown in Appendix A (Equation A-12) to account for air leakage into filtered environmental enclosures. This model, shown below, describes cab penetration (*Pen*) in terms of fractional intake filter efficiency (η_f), intake air leakage (*l*), inside concentration (*x*) and outside concentration (*c*) at steady-state cab conditions. Intake air leakage (*l*) is defined as the proportion of cab intake airflow that bypassed the filter media and is unfiltered.

$$Pen = \frac{x}{c} = 1 - \eta_f + l\eta_f \qquad (A-12)$$

Results from the initial laboratory experiments indicated that this steady-state atmospheric gas testing methodology can detect the amount of air leakage into filtered environmental enclosures, but additional research was needed to refine the level of accuracy that can be achieved with the instrumentation and test filters [Organiscak and Schmitz 2006].

Additional research was conducted to examine several grades of carbon dioxide instrumentation and to refine the leakage testing methodology for improving measurement accuracy. Tests were conducted to examine the operational differences between three instruments, having diverse performance specifications. A multiple-location cab sampling methodology was also developed for a single-instrument evaluation of cabs. The single-instrument sampling methodology accentuates instrument precision characteristics for leak testing cabs while diminishing calibration inaccuracies or biases associated with using multiple instruments. This paper describes the tests conducted with the carbon dioxide instruments and the single-instrument, multiple-location cab sampling methodology development.

Examination of CO_2 Instruments

Three different types of infrared-sensing carbon dioxide instruments were examined for their suitability for the proposed leak test. The instruments chosen were conveniently available to NIOSH and CAF for testing. These instruments included Telaire 7001 carbon dioxide monitors (Goleta, Calif.), Vaisala GM70 carbon dioxide meters (Helsinki, Finland), and a Sable CA-10a carbon dioxide analyzer (Sable Systems International, Las Vegas, Nev.). Table 1 shows the manufacturer specifications and approximate cost for each of these products. These only represent a few of the many instruments available from other manufacturers, but indicate the performance range that can exist between different grades of instruments. The Telaire and Vaisala were lower-cost, hand-held passive sampling instruments, and the Sable was a higher-cost, portable benchtop active sampling instrument. NIOSH had two Vaisala and two Telaire instruments available for testing, and these instruments are referred to specifically as Vaisala 1, Vaisala 2, Telaire NIOSH1, and Telaire NIOSH2 throughout this paper. CAF had one Sable and two Telaire instruments available for testing, and these instruments are referred to specifically as Sable, Telaire CAF1, and Telaire CAF2 throughout this paper.

According to Table 1, these instruments had different specified levels of accuracy, response times, and user calibration capabilities. The Telaire was the least accurate instrument ($\pm$ 50 ppm or $\pm$ 5% of reading) with a response time specified as less than 60 seconds. It could only be zeroed with nitrogen or one-point calibrated to a known gas concentration by the user. The Vaisala was the second most accurate instrument ($< \pm$ 20 ppm + 2% of reading) with a 30 second response time, and the Sable was the most accurate instrument (better than 1% of calibrated span) with a ½ second response time. Both these instruments could be two-point calibrated (zeroed and spanned) by the user. Because the Sable had 11 selectable measurement ranges, it was the most flexible in that it could be spanned over a narrow or wide CO_2 concentration range.

Table 1. Carbon dioxide instrument specifications

Instrument Specifications	Telaire Model 7001	Vaisala Model GM70	Sable Model CA-10a
Sample method	Passive diffusion	Passive diffusion or pump aspiration	Active sample pump 400 ml/min
Measurement range	0–10,000 ppm display 0–4,000 ppm analog output	GMP 222 probe 0–2,000 ppm	11 Selectable ranges Used ranges at or under 0–0.20% or 2,000 ppm
Data logging capability	0–2 V analog output	Internal memory 2,700 points	0–5 V analog output
Accuracy	The larger of ± 50 ppm or ± 5% of reading	< ± [20 ppm + 2% of reading]	Better than 1% of calibrated span
Resolution	1 ppm	10 ppm	1 ppm up to 2,000 ppm
User calibration	Zero or one point	Two-point span	Two-point span
Pressure compensation	User input of elevation	User input of barometric pressure	Instrument adjusted barometric pressure
Temperature compensation	None	User input of temperature	Software adjusted
Response time	< 60 seconds	30 seconds	0.5 second
Warm up time	< 60 seconds	15 minutes	Several hours
Cost	Approx. $500	Approx. $2,000	Approx. $6,500

Experimental Test Methods

All of these instruments were tested against three known calibration gases and comparatively tested side-by-side in an enclosure at various CO_2 concentration levels.[3] They were not evaluated under the NIOSH Guidelines for Air Sampling Analytical Method Development and Evaluation [NIOSH 1995]. These tests were performed at CAF's test facility in Defiance, Iowa. The three calibration or reference gases used for instrument calibration, validation, and testing were 0 ppm (pure nitrogen), 20.1 ppm (CO_2 and nitrogen mix), and 400 ppm (CO_2 and nitrogen mix). Before these instruments were tested, NIOSH sent their Telaire and Vaisala instruments back to the manufacturers for calibration; these instruments were initially tested as calibrated by the manufacturer. CAF purchased the Sable and one of their Telaire instruments before testing. CAF's older Telaire was rezeroed, but the newly purchased Telaire was initially tested as calibrated by the manufacturer. The Sable was zeroed with nitrogen and spanned with the 400 ppm reference gas.

---

[3] The instrument comparisons made in this study are intended to be illustrative of such categories of instruments available for cab/leakage testing, but should not be considered definitive with respect to their performance or reliability in other environmental or ventilation monitoring applications.

These three types of instruments were used to measure the known gas concentrations flowing from the reference gas tanks. The Sable actively drew gas at 400 ml/min through sample tubing from the larger opening side of a syringe reservoir packed off with permeable facial tissue that was positively pressurized by 500 ml/min of regulated reference gas flow from the tank. This procedure avoided over pressurizing the Sable's sampling pump from the highly pressurized gas cylinders. Reference gases were regulated at a positive pressure of 7 psig from the tank into the calibration ports of the Telaire or Vaisala instruments. Carbon dioxide (CO_2) concentration averages of 15 seconds were recorded over 5-minute sampling periods for each instrument and reference gas. The Sable and Telaire analog outputs were recorded by a Telog 3307 multichannel data acquisition system (Telog Instruments, Inc., Victor, N.Y.). The Vaisala measurements were stored in the internal memory of the instrument.

After the reference gas measurements were completed, instrument tests were conducted under more realistic atmospheric sampling conditions inside CAF's experimental cab test stand. Figure 1 shows the cab test stand setup for instrument comparisons. The cab test stand is a simulated plywood cab enclosure with a known volume of 52.5 ft^3. It does not represent any particular cab, but a physical model that can have its filter, airflow, and pressure characteristics controlled during laboratory testing. The four Telaire and two Vaisala passive diffusion sampling instruments were placed side-by-side inside the center of the cab test stand. The Sable was placed outside the enclosure with the inlet of its sampling hose place at the center of the enclosure near the other instruments. The interior of the cab test stand was initially sampled at atmospheric CO_2 concentrations just after its entry door was closed. The CO_2 concentrations were sequentially stepped to lower concentrations by turning the air filtration system on and off during instrument testing. A carbon dioxide panel filter (18-in x 11-in x 2.19-in thick) was fitted on the intake air filtration system to lower CO_2 concentrations inside the enclosure. CO_2 concentration averages of 15 seconds were again data logged during testing, and instrument comparisons were only made during stable 5-min concentration periods with the filtration system turned off. A small 6-in-diameter table fan was placed on a small bench inside the back corner of the enclosure to ensure adequate mixing and movement of inside air for the passive diffusion instruments when the filtration system was off.

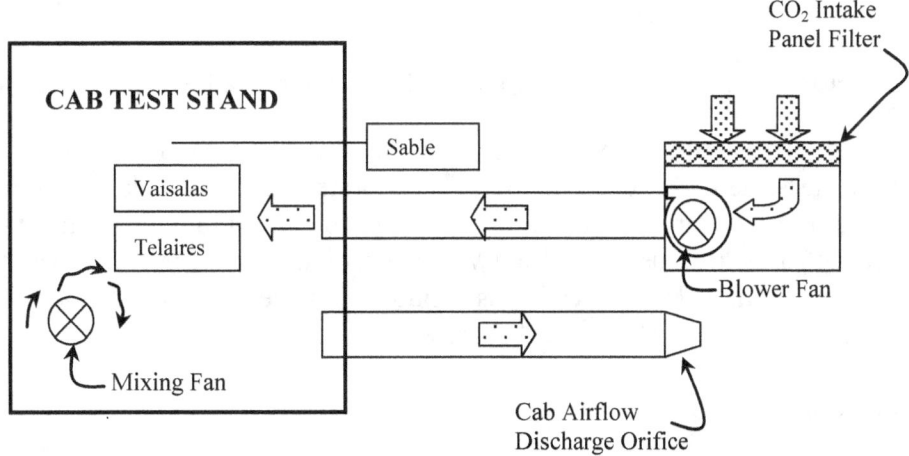

Figure 1. CAF cab test stand setup for comparative instrument testing.

Two series of instrument tests were made inside CAF's experimental cab test stand. Test Series 1 was conducted with the instruments after the initial reference gas testing, and Test Series 2 was conducted after the instruments were recalibrated (Sable and Vaisala instruments) or rezeroed (Telaire instruments) by the user. These instruments were further re-evaluated as previously described with the reference gases after the second series of cab tests.

A final examination of these instruments was also conducted inside a John Deere 7820 tractor cab. Figure 2 shows the tractor cab setup for instrument comparisons. The tractor cab intake air duct was fitted with a 3-in-diameter tee and straight section of PVC pipe leading to the intake filter box fitted with a carbon dioxide panel filter (18-in x 11-in x 2.19-in thick). The perpendicular opening of the tee was used for regulating outside air leakage into the filtered intake air flowing into the cab. The tee was completely opened to the atmosphere at the start of the instrument testing with the tee opening sequentially reduced with PVC fittings to decrease outside air leakage around the filter, thus lowering the CO_2 concentrations inside the cab. The Sable, one Vaisala, and four Telaire instruments were placed side-by-side on the operator's seat in the tractor cab. Concentration averages of 15 seconds were again data logged during testing, and instrument measurements were only examined during the stable 5-min concentration periods under various leakage conditions at the tee.

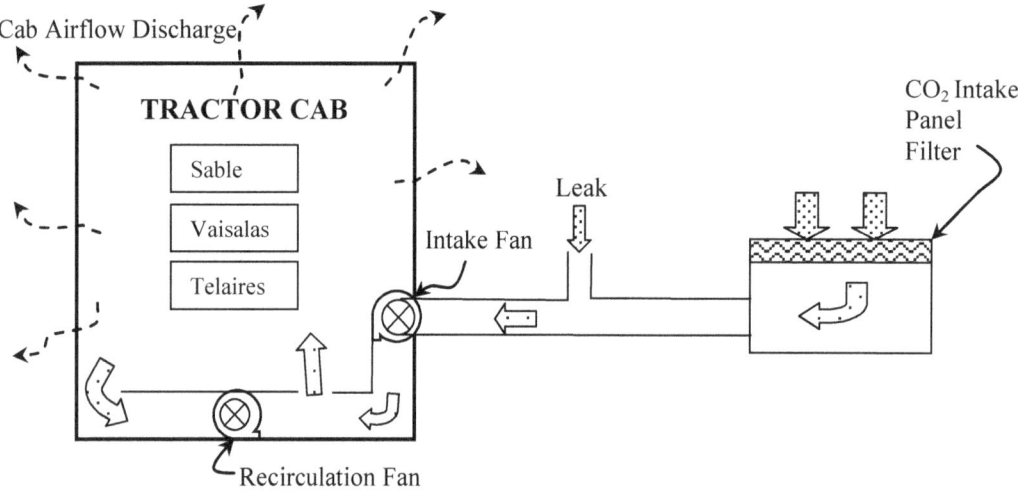

Figure 2. John Deere tractor cab setup for comparative instrument testing.

Two series of instrument tests were made inside the John Deere tractor cab. Test Series 1 was conducted after the Sable and one Vaisala were recalibrated, and four Telaires were rezeroed by the user. Several of the Telaires had to be rezeroed multiple times to improve their agreement with the reference gases. Because one of the Vaisalas would not recalibrate, it was excluded from the tractor cab testing. Test Series 2 was conducted inside the cab after the Vaisala and two Telaires were single-point calibrated to 20 ppm.

Instrument Test Results

The instrument concentration averages and standard deviations initially measured during the 5-minute periods with the different reference gas concentrations are shown in Table 2. As can be seen from this table, and consistent with manufacturer representations, the Sable instrument appeared to be the most accurate and precise instrument. The Sable averaged within 2 ppm of the reference gas concentrations and had a standard deviation below 1 ppm. The accuracy and precision of the other instruments were noticeably less and inconsistent as compared to the Sable instrument. The factory-calibrated Telaire and Vaisala instruments showed both higher and lower measurement biases as compared to the reference gases.

Table 2. Initial instrument comparisons with reference gases.

Instrument	Avg. of calibration gas concentrations			Std. deviation of calibration gas concentrations			Pooled Std. deviation
	0 ppm	20.1 ppm	400 ppm	0 ppm	20.1 ppm	400 ppm	All gases
Sable CA-10a	2	21	400	0.9	0.4	0.5	0.7
Telaire CAF1	2	29	395	1.8	3.0	3.2	2.7
†Telaire CAF2	0	0	372	0.0	1.2	5.2	3.1
†Telaire NIOSH1	11	29	418	3.8	3.6	7.1	5.1
†Telaire NIOSH2	4	32	428	0.7	1.0	28.2	16.3
†Vaisala 1	11	21	402	2.2	2.2	5.2	3.5
†Vaisala 2	-23	-18	377	4.7	16.2	4.9	10.1

†Instrument measurements with manufacturer calibration.

Instrument tests made inside CAF's experimental cab test stand following the reference gas measurements are shown in Figure 3 and Table 3. Figure 3 shows the Telaire and Vaisala concentration measurements as compared to the Sable for Test Series 1. As can be seen in Figure 3, the Telaire and Vaisala instruments exhibited linear relationships with the Sable. A unity line is also shown on Figure 3 and illustrates that the Telaire and Vaisala instruments exhibited either a higher or lower measurement bias as compared to the Sable instrument. Similar linear relationships with changed measurement biases were observed for the Test Series 2 after the Telaire instruments were rezeroed, and the Vaisala and Sable instruments were recalibrated.

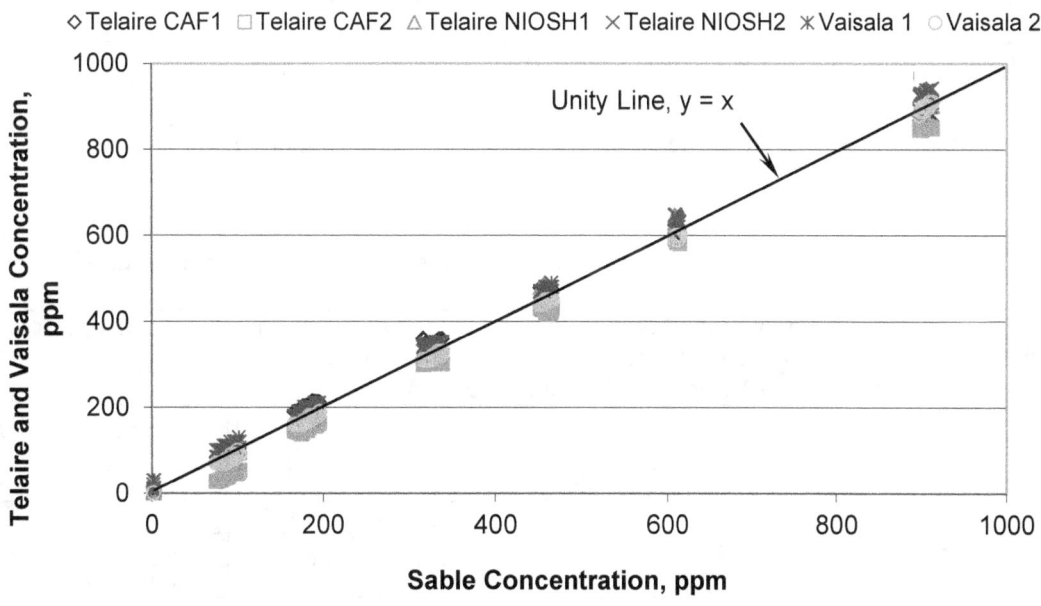

Figure 3. Instrument measurements inside CAF cab test stand during test series 1.

Table 3 shows the linear regression analyses of the Telaire and Vaisala measurements with respect to the Sable inside the CAF's experimental cab test stand. These regression results showed a good linear fit of the data with regression slopes (b) and coefficient of determinations (r^2) near 1. However, all of these instruments, with the exception of the Telaire NIOSH1 instrument during Test Series 1, were not considered to be equivalent to the Sable because their regression intercepts (a) and slopes (b) were significantly different from 0 and 1, respectively, at the 95% confidence level. Regression intercepts ranged from -21.9 ppm to 22.9 ppm for Test Series 1 and ranged from -27.2 ppm to 22.0 ppm for Test Series 2. Noticeable changes to the regression intercepts were observed for most of the instruments between the two test series after recalibration. Finally as shown in table 3, the Telaire instruments exhibited consistently higher regression standard deviations when compared to the Vaisala instruments.

Table 3. Linear regression parameters of Telaire and Vaisala instruments as compared to the Sable instrument inside the CAF test stand.

Instrument	Test Series 1 inside CAF test stand				Test Series 2 inside CAF test stand			
	Intercept, a	Slope, b	r^2	s	Intercept, a	Slope, b	r^2	s
Telaire CAF1	22.9	0.96	0.997	14.2	21.9	0.97	0.996	16.4
Telaire CAF2	-21.9	0.97	0.997	14.9	22.0	0.99	0.998	11.6
Telaire NIOSH1	*1.3	*1.00	0.998	12.1	-19.4	0.97	0.997	15.3
Telaire NIOSH2	3.0	1.03	1.000	6.6	-27.2	0.99	0.997	14.7
Vaisala 1	22.3	0.97	1.000	4.8	2.1	1.03	1.000	5.0
Vaisala 2	-11.1	1.00	1.000	4.3	-7.3	1.01	1.000	4.7

Note: Linear Regression Model is y = a + bx, where y is the Telaire or Vaisala instrument concentration and x is the Sable concentration.
*The regression intercept (a) and slope (b) parameters were not significantly different from 0 and 1, respectively, at the 95% confidence level.

Further reference gas evaluations of these instruments were conducted after CAF test stand comparisons; these results are shown in Table 4. As can be seen from this table, the Sable instrument appeared to maintain its accuracy and precision. The Sable again averaged within 2 ppm of the reference gas concentrations and had a standard deviation below 1 ppm. The accuracy and precision of the Telaire and Vaisala instruments were noticeably less and inconsistent. Their accuracy evidently changed from the first reference gas evaluations in Table 1 after rezeroing/recalibration and usage.

Table 4. Instrument comparisons with reference gases after user recalibration.

Instrument	Avg. of calibration gas concentrations			Std. deviation of calibration gas concentrations			Pooled Std. deviation
	0 ppm	20.1 ppm	400 ppm	0 ppm	20.1 ppm	400 ppm	All gases
Sable CA-10a	1	20	398	0.4	0.4	0.6	0.5
Telaire CAF1	19	61	443	3.9	5.0	7.7	5.8
Telaire CAF2	32	57	435	3.5	2.3	1.9	2.6
Telaire NIOSH1	2	52	402	1.5	5.1	1.6	3.2
Telaire NIOSH2	4	0	377	4.6	0.0	4.2	3.6
Vaisala 1	3	21	390	4.4	3.1	0.0	3.1
Vaisala 2	0	1	359	0.0	2.2	3.1	2.2

The additional instrument tests conducted inside a John Deere 7820 tractor cab showed similar linear measurement characteristics with instrumental biases as observed in the CAF cab test stand. Before Test Series 1 inside the tractor cab, the instruments were warmed-up, rezeroed, and recalibrated. The Vaisala 1 instrument was removed from the tractor cab testing because it was malfunctioning and would not recalibrate. Figure 4 shows the Telaire and Vaisala instrument measurements as compared with the Sable instrument made during Test Series 1 inside the tractor cab. As previously observed in the CAF test stand comparisons, the Telaire and Vaisala concentration measurements exhibited linear relationships as compared to the Sable. Before Test Series 2, only the Telaire CAF2, Telaire NIOSH2, and Vaisala 2 were single-point calibrated to the 20.1 ppm reference gas concentration.

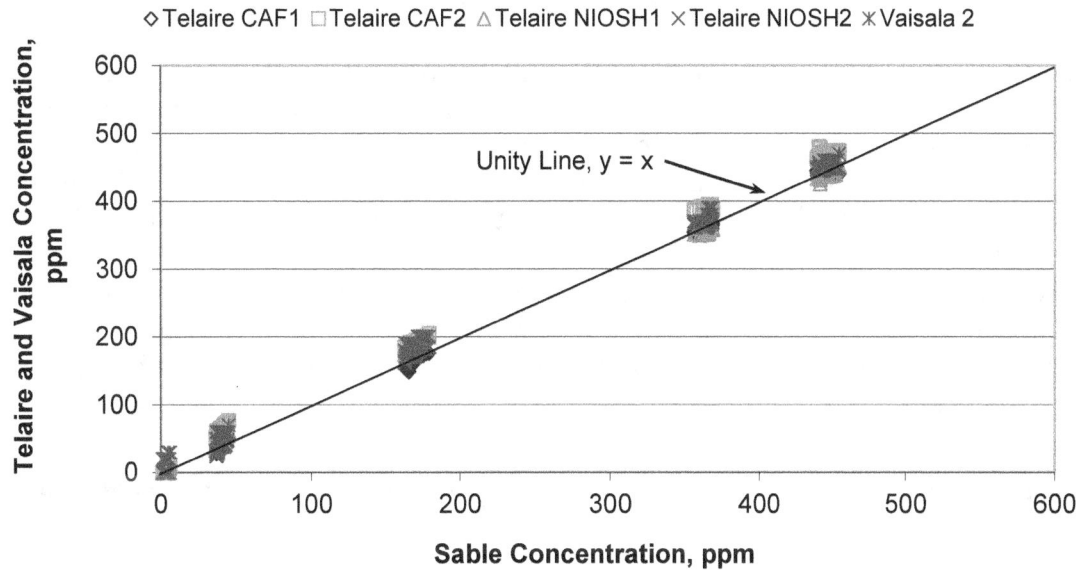

Figure 4. Instrument measurements inside John Deere tractor cab during test series 1.

Table 5 shows the linear regression parameters fitted for the hand-held Telaire and Vaisala instruments as compared to the Sable instrument for Test Series 1 and 2 inside the tractor cab. These regression results once more showed a good linear fit of the data with regression slopes (b) and coefficient of determinations (r^2) near 1. These instruments were again not considered to be equivalent to the Sable because all of their regression intercepts (a) and slopes (b) were significantly different from 0 and 1, respectively, at the 95% confidence level with the exception of the regression slope (b) for the Telaire NIOSH1 instrument during Test Series 1. The most noteworthy differences between these instruments were, again, their regression intercepts (a) with several instrumental changes observed between Test Series 1 and 2 inside the tractor cab, regardless of recalibration status.

Table 5. Linear regression parameters of Telaire and Vaisala instruments as compared to the Sable instrument inside the John Deere tractor cab.

Instrument	Test Series 1 inside John Deere tractor cab				Test Series 2 inside John Deere tractor cab			
	Intercept, a	Slope, b	r^2	s	Intercept, a	Slope, b	r^2	s
Telaire CAF1	3.4	0.98	0.998	8.6	16.1	1.03	0.998	6.1
†Telaire CAF2	-2.3	1.02	0.999	4.8	-4.6	1.03	0.999	4.8
Telaire NIOSH1	-3.6	*1.00	0.999	5.3	-8.4	1.04	0.999	6.8
†Telaire NIOSH2	11.4	1.04	0.998	8.0	-3.2	1.03	0.998	7.1
†Vaisala 2	21.0	0.98	0.999	4.7	19.6	1.00	0.999	4.1

Note: Linear Regression Model is y = a + bx; where y is the instrument conc. and x is the Sable conc.
*The regression slope (b) parameter was not significantly different from 1 at the 95% confidence level.
†These instruments were single-point calibrated at 20.1 ppm of reference gas for Test Series 2.

These test results have indicated that the Sable, Telaires, and Vaisalas are linear instruments with respect to CO_2 concentrations. However, and consistent with manufacturer representations, the Telaire and Vaisala instruments were found to be less accurate and precise. During our testing, the less expensive instruments were also more difficult to accurately recalibrate. It was also apparent from this testing that all of these instruments can drift and would need daily checks or calibration to ensure their accuracy. Inaccuracy or biases between multiple sampling instruments increases their measurement error of cab penetration and/or cab leakage. Because maintaining instrument accuracy for field testing cabs can be challenging, a single-instrument, cab testing methodology was developed to potentially do away with instrument bias errors and field calibrations.

Development of New Cab Leakage Sampling Methodology

The linear and relatively precise characteristics of the CO_2 instruments tested may be used to provide accurate cab leakage field measurements without frequent instrument calibration. One way this may be achieved is by measuring relative CO_2 concentration differences inside the cab filtration system by using the same instrument at or near steady-state conditions. In order to accomplish this type of leakage assessment, Equation A-12, previously presented above, is reformulated into leakage Equation A-17 in Appendix A and is expressed in CO_2 concentrations as measured inside the cab, after the intake filter, and outside the cab under steady-state cab test conditions. This reformulated leakage Equation A-17 is shown below.

$$l = \frac{x-i}{c-i} \qquad (A\text{-}17)$$

Where: x = Inside cab concentration (ppm)

c = Outside cab concentration (ppm)

i = Immediate concentration after the intake filter (ppm)

l = Proportion of intake air leakage into the cab

And: $x \geq i$, $c > i$, and $c \geq x$

Given the linear and relatively precise nature of the CO_2 instruments previously tested, it is assumed that instrument inaccuracies can be negated when using the single-instrument sampling methodology of Equation A-17. An active CO_2 instrument like the Sable would be more suitable to take samples from tubing at these multiple locations on the tractor cab. To study this sampling methodology, cab filtration system leakage experiments were conducted on CAF's laboratory test stand.

Experimental Test Methods

Air leakage testing with the new single-instrument, multiple-location sampling methodology was conducted on CAF's experimental test stand. Figure 5 shows the schematic of the cab air leakage test setup. Two CO_2 intake panel filters (18-in x 11-in x 2.19-in thick) were stacked to ensure negligible filter penetration for the leak testing experiments. Because the filtration system was well-sealed with gaskets, silicon, and duct seal compound (putty), the only air leakage source into the system was presumed to be the controlled and quantifiable leak downstream of the filter panels. A TSI Model 4040 Thermal Mass Flowmeter (TSI, Inc., Shoreview, Minn.) measured the controlled air leakage around the intake filter, and its data was continuously recorded on a Telog data acquisition system (Telog Instruments, Inc., Victor, N.Y.). A TSI Model 8345 VELOCICALC Hot Wire Anemometer (TSI, Inc., Shoreview, Minn.) was used to measure the cab intake airflow (3-in-diameter pipe centerline measurement) to the cab test stand at the beginning and end of each leak test condition. These centerline airflow measurements were multiplied by a 0.85 centerline velocity factor to determine the average airflow throughout the cross section of the PVC pipe for the high Reynolds Number turbulent flow condition [Knudsen and Katz 1958]. The portion of air leakage around the intake filter (l) could be determined by dividing the average leakage airflow quantity by the average cab intake airflow quantity for the test.

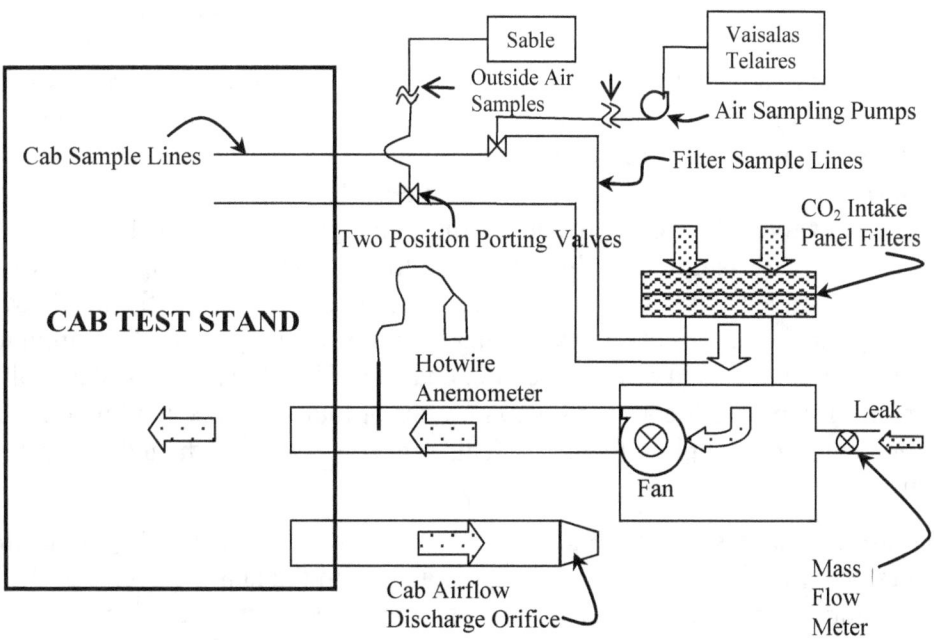

Figure 5. CAF cab test stand setup for leakage experiments.

Three leakage levels were evaluated in the laboratory. The three leakage levels studied were no leak, a ¼-in-diameter orifice leak, and a ¾-in-diameter orifice leak. The fan speed was electronically adjusted by a variable voltage source to try and maintain about 25 cfm of intake air into the cab for the three leakage conditions. This yielded approximate leakage percentages of 0%, 5.5%, and 11.5% during laboratory testing.

One Sable, two Vaisala, and two Telaire instruments were used to sequentially sample CO_2 concentrations: (1) inside the cab, (2) immediately after the intake filter, and (3) outside the cab at or near-steady operating conditions. To conduct this type of sampling procedure, the instruments were located outside the unoccupied cab test stand, and the air samples were actively drawn from the multiple locations through 1/8-in inside diameter Tygon tubing. The Sable is an active sampling instrument with its own air pump and was found to be ideal for this type of sampling procedure. Because the Vaisalas and Telaires are passive sampling instruments without pumps, SKC Model 224-PCXR4 universal air sampling pumps (SKC, Eighty Four, Pa.), operating at 1.0 L/min, were used to actively pump air from the sampling locations to the Vaisala and Telaire instruments. The sampled air, drawn with SKC sampling pumps, was pumped into the Vaisala instruments' calibration hood or into the Telaire instruments' calibration port. Table 1 specifies that the Vaisala instruments can be aspirated with a sampling pump, and that the Telaire instruments are primarily passive samplers. However, the Telaire instruments were also pump aspirated in these experiments to examine their performance for this type of sampling procedure. Instrument comparisons with calibration gases before testing showed that the Sable operated within a few ppm of the known gas concentrations, although the Vaisala and Telaire instruments varied up to 40 ppm from the known gas concentrations.

Two initial series of air leakage tests were conducted with the CO_2 instruments. For Test Series 1 the Sable and one of the Vaisalas simultaneously sampled the same cab locations with the other Vasiala sampling a different cab location. For Test Series 2, the Sable, one Telaire, and one Vaisala simultaneously sampled the same cab locations with another Telaire sampling a different cab location. The Sable and one of the other instruments initially sampled the inside cab concentrations to detect when they reached their lowest stable level after the cab test stand door was closed (at or near steady-state conditions). The CO_2 concentration measurements for this leak test methodology started only after the cab interior was stable. All the instruments were sequentially rotated through the three sampling locations twice. Instrument air samples drawn from inside the cab were successively switched to immediately after the intake filter by using two-position porting valves (Dwyer Instruments, Inc. Michigan City, Ind.). The outside cab location was consecutively sampled by disconnecting the instruments' sampling hoses from the two-position porting valves and later reconnecting these hoses to return to the inside cab location. The CO_2 concentrations at each sampling location were measured for 5 minutes after stabilizing for a 5-minute period between location changes. Thus, 10 minutes of sampling time was required at each sampling location with 60 minutes of total time needed for completing two rotations through the three sampling locations. This procedure was conducted under the three leakage conditions.

Given that these initial tests were conducted during the winter season, CO_2 concentrations inside the laboratory were found to fluctuate appreciably with the heating cycles of the building. Additional laboratory tests were conducted in late spring to re-examine cab leakage measurements under more stable CO_2 test conditions without the heating cycle interferences. During these tests a refined sampling methodology was devised to help average out some of the outside CO_2 concentration variation effects on cab leakage measurements. This methodology shortened the instrument stabilization and sampling times at each location and allowed an increase in the sampling frequency that can be made at each location for a given test time period. The new refined methodology reduced the instrument stabilization time to 2 minutes and the CO_2 concentration measurement interval to 3 minutes. This requires 5 minutes of sampling at each location and 15 minutes for instrument rotation through the three sampling locations. Thus, the frequency of sampling at each location could be increased for a given test time period. Two minutes of stability time was considered adequate to cover the few seconds it would take for air samples to reach the instruments through nearly 7 ft of 1/8-in-diameter sample tubing and up to 1 minute of response time for the Telaire instruments.

This new refined test methodology was evaluated for the three leakage conditions previously tested. However, during these tests the Sable, two Vaisalas, and one Telaire were simultaneously sampled inside the cab, immediately after the intake filter, and outside the cab for three complete rotations. The leak test measurements again started after the inside cab concentrations stabilized and the cab was operating at or near steady-state conditions. Three sampling rotations were conducted at all the sampling locations for a total sampling time of 45 minutes during each leakage condition. All the leakage conditions were replicated with this methodology during the two test series. Because these tests were conducted in late spring, the laboratory concentrations were reasonably steady due to less heating cycles and better laboratory ventilation.

Finally, the use of the refined leak testing methodology was demonstrated on a John Deere 7820 tractor cab as shown in Figure 6. Because the filtration system leakage area(s) into this tractor cab were unknown and air leakage could not be quantified by airflow measurements as previously conducted in the laboratory test setup, only the refined CO_2 leak testing methodology was used. For this test a Clean Air Filter JD60R cylindrical CO_2 test filter cartridge was constructed with two sealed sampling tubes placed through the filter housing to the immediate downstream side of the filter. The intake test filter was inserted into the tractor, and the Sable and Vaisala 1 instruments were used to measure concentrations inside the cab, immediately after the filter, and outside the cab. The instruments were simultaneously rotated three times through these sampling locations as just previously described for the refined test methodology.

Figure 6. Leak testing the John Deere tractor cab. (Photo courtesy of Clean Air Filter).

Test Methodology Results

Air leakage was calculated from the air quantity measurements made during the laboratory tests and compared to the CO_2 determined leakage as measured by each of the instruments used during the tests. The average mass flow meter quantity during the leakage condition was divided by the average hot wire intake air quantity measured before and after each leakage test. The air leakage measured by each CO_2 instrument was determined from using the sequential concentrations measured inside the cab, immediately after the intake filter, and outside the cab in Equation A-17. The proportional leakages measured are expressed as percentages throughout this paper.

The initial leakage methodology test results are shown in Figures 7 and 8. The data for both initial test series are shown in Appendix B. As illustrated by Test Series 1 in Figure 7, the outside cab test stand laboratory CO_2 concentrations noticeably varied during these initial leakage tests. The outside concentration ranged on average by 76 ppm between sampling rotations for each leakage condition tested. Because the leakage Equation A-17 assumes steady-state conditions, these concentration variations were reflected in the CO_2 leakage variations calculated and shown in Figure 8 for both Test Series 1 and 2. Figure 8 illustrates the notable variation in CO_2 leakage determinations around a unity line as compared to the proportional air leakage quantities measured. The notable CO_2 variations in some of the no-leak tests yielded negative leakages.

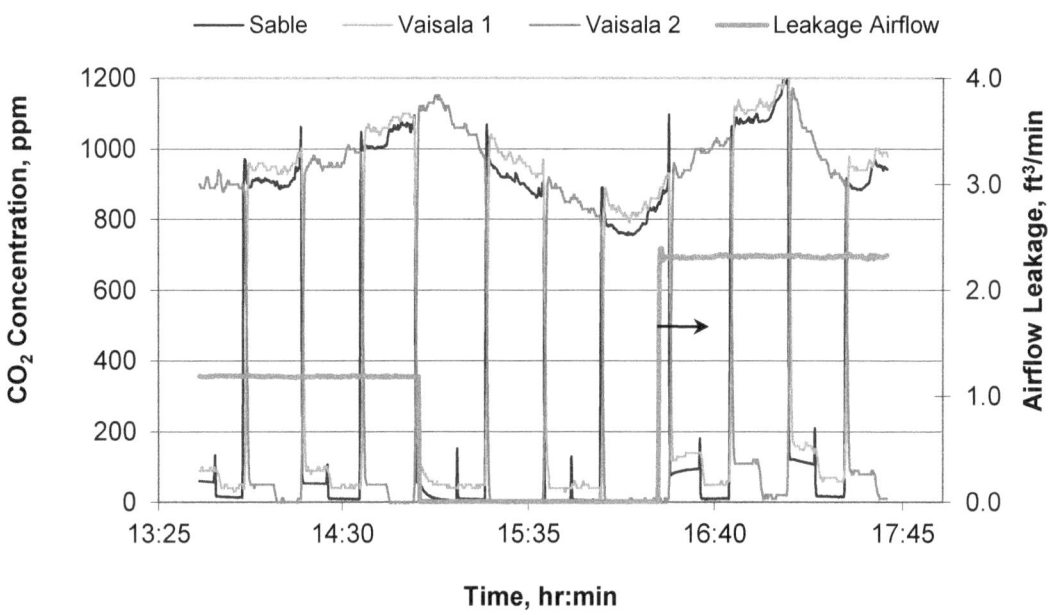

Figure 7. CO_2 concentrations for the first series of initial leakage tests.

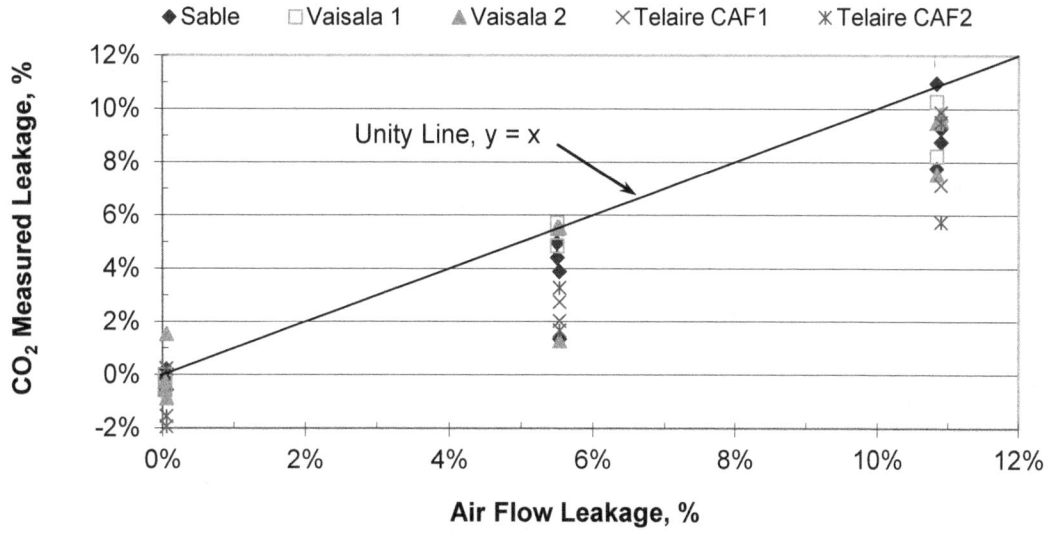

Figure 8. Initial leakage test measurements made on the CAF test stand.

The refined leakage test methodology results are shown in Figures 9 and 10. The data for both refined test series are shown in Appendix C. As indicated in Figure 9, the outside cab test stand laboratory concentrations were more uniform during these leakage tests. The outside concentration ranged on average by 17 ppm between sampling rotations for each leakage condition tested. The steadier concentrations were reflected in the leakage comparisons shown in Figure 10 for both series of CO_2 instrument tests. Figure 10 illustrates less variation in Sable and Vaisala CO_2 leakage determinations around a unity line as compared with Figure 8 for the initial tests. The Telaire leakage measurements continued to show wide variations which could be a reflection of the Telaire's slower response to concentration changes and/or larger variations in concentration measurements as previously observed during instrument testing.

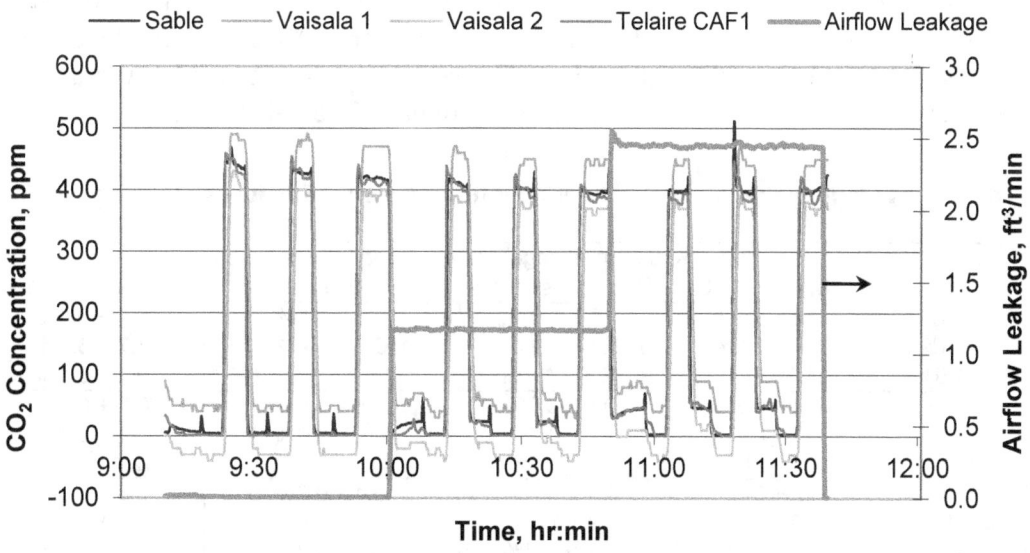

Figure 9. Typical CO_2 concentrations during refined leakage tests.

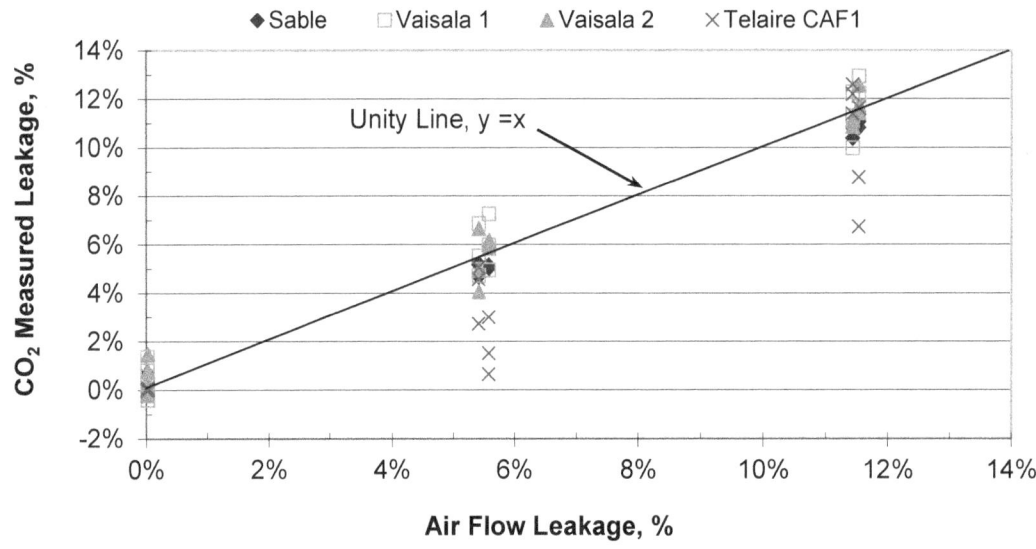

Figure 10. Refined leakage test measurements made on CAF test stand.

Table 6 shows the averages and 95% confidence intervals for both test series of the refined CO_2 leakage sampling methodology. Each test series has three complete rotations between the sampling locations for each leakage condition. As can be seen from this table, the Sable instrument yielded the most accurate and precise CO_2 leakage measurements. The Sable leakages were within 0.7% of the air quantity leakage measurements with confidence intervals up to 0.7%. When both test series data (replicates) were averaged together, the Sable leakage confidence intervals decreased to $\leq 0.3\%$. Leakages determined from the Vaisalas showed reasonable agreement with the air quantity measurements of leakage. The Vaisala leakages were usually within 1.0% of the air quantity leakage measurements with confidence intervals up to 3.3%. When both test series data (replicates) were averaged together, the Vaisala confidence intervals decreased to $\leq 1.1\%$. The Telaire instrument had the most notable disagreement with the air leakage measurements. The Telaire leakages were greater than 1.0% of the air quantity leakage measurements for about half of the tests with confidence intervals up to 6.1%. Averaging both tests series data (replicates) reduced the Telaire's confidence intervals to $\leq 2.4\%$.

Table 6. Average leakage and confidence intervals for CAF test stand experiments

Test series	Airflow leakage, %	CO_2 Instrument leakage averages and 95% confidence intervals			
		Sable CA-10a	Vaisala 1	Vaisala 2	Telaire CAF1
1	0.0%	0.2 ± 0.7%	1.0 ± 1.0%	0.0 ± 0.8%	0.0 ± 0.0%
	5.6%	5.1 ± 0.2%	6.1 ± 2.9%	6.0 ± 0.4%	1.7 ± 3.0%
	11.5%	11.1 ± 0.7%	12.4 ± 1.1%	11.9 ± 1.4%	9.1 ± 6.1%
2	0.0%	0.4 ± 0.7%	-0.1 ± 0.7%	0.8 ± 1.6%	0.0 ± 0.1%
	5.4%	5.0 ± 0.7%	5.7 ± 2.7%	5.2 ± 3.3%	4.1 ± 3.0%
	11.4%	10.7 ± 0.6%	11.2 ± 2.9%	11.0 ± 0.4%	12.1 ± 1.5%
1&2	0.0%	0.3 ± 0.3%	0.4 ± 0.7%	0.4 ± 0.7%	0.0 ± 0.0%
	5.5%	5.0 ± 0.2%	5.9 ± 1.1%	5.6 ± 1.0%	2.9 ± 1.8%
	11.5%	10.9 ± 0.3%	11.8 ± 1.1%	11.5 ± 0.6%	10.6 ± 2.4%

Table 7 shows the leakage testing methodology demonstrated on the John Deere 7820 tractor cab. The intake filter cartridge housing has a limited area for inserting a couple of air sampling tubes to the downstream side of the intake filter. Only the Sable and Vaisala 1 instruments were used simultaneously during this test. The pressure differential measured across the intake filter was 0.74 in of water gauge during cab testing. This pressure differential was equivalent to 37 ft^3/min of airflow through the filter as measured on the CAF cab test stand. As can be seen in Table 7, the intake air filter did not have to remove all of the CO_2 to detect and measure outside air leakage into the cab. The Sable instrument appeared to provide the most reliable leakage measurements. The Sable instrument detected a 1.1% cab filtration system leak with a 95 confidence interval of 0.3%. The Vaisala instrument detected a 1.5% leak with a confidence interval of 2.3%. These test results demonstrate that the multiple-sampling-location methodology is a viable cab leakage test method when using a specially constructed intake test filter to allow for CO_2 concentration sampling immediately downstream of the filter. Because cab filtration system leakage area(s) are usually unknown, cab leakage measurements determined from airflow quantities would be impractical.

Table 7. Average leakage and confidence intervals for John Deere 7820 tractor cab

CO_2 instrument	Cab concentration. (ppm)	Filter concentration (ppm)	Outside concentration (ppm)	Cab leakage (%)	Average cab leakage (%)
Sable CA-10a	19	14	420	1.2%	1.1 ± 0.3%
	18	13	405	1.2%	
	17	13	429	1.0%	
Vaisala 1	47	43	441	0.8%	1.5 ± 2.3%
	50	46	432	1.1%	
	52	42	438	2.5%	

Discussion of Cab Leakage Measurement Error

The above laboratory experiments showed that both instrument precision and outside cab concentration variations can have an effect on the cab leakage measurement errors. Although using the same instrument for cab leakage testing is expected to eliminate instrument bias or accuracy errors, random and/or experimental errors are present in the testing. Random errors are primarily a result of instrument precision, whereas experimental errors are a result of systematic deviations from steady-state test conditions. In order to examine the size of these errors, a propagation of error analysis was conducted for the cab leakage measurement methodology [Bevington 1969]. The uncertainty or relative standard deviation (RSD) of the leakage methodology was derived in Appendix D and is shown below in Equation D-6. The relative standard deviation of this leakage methodology (RSD_l) is the standard deviation (s_l) of proportional leakage measurements divided by its mean (l). On the right side of this equation, the first three terms tend to reflect the random errors during testing, whereas the last three terms tend to reflect the experimental systematic errors of testing with unstable outside concentrations. At steady-state cab test conditions, the last three systematic error terms in the equation become zero, because the measured concentrations are uncorrelated and have zero covariances.

$$RSD_l = \frac{s_l}{l} = \sqrt{\begin{array}{c} s_x^2 \dfrac{1}{(x-i)^2} + s_i^2 \dfrac{(c-x)^2}{(c-i)^2(x-i)^2} + s_c^2 \dfrac{1}{(c-i)^2} - 2s_{xi}^2 \dfrac{(c-x)}{(c-i)(x-i)^2} \\ - 2s_{xc}^2 \dfrac{1}{(c-i)(x-i)^2} + 2s_{ci}^2 \dfrac{(c-x)}{(c-i)^2(x-i)} \end{array}}$$ (D-6)

where l = mean of proportional leakage,
s_l = standard deviation of proportional leakage,
RSD_l = relative standard deviation of leakage estimate, s_l/l,
x = mean cab concentration,
s_x = standard deviation of cab concentration,
i = mean downstream filter concentration,
s_i = standard deviation of downstream filter concentration,
c = mean outside cab concentration,
s_c = standard deviation of outside cab concentration,
s_{xi}^2 = covariance between x and i,
s_{xc}^2 = covariance between x and c,
s_{ci}^2 = covariance between c and i.

This equation can be useful for estimating the relative standard deviation of the CO_2 cab leakage testing methodology given the means, standard deviations, and covariances of the concentrations. It was used to examine the primary effects of instrument precision (i.e., random error) and intake filter efficiency on cab leakage measurement error. In this analysis the systematic error terms were neglected, assuming steady-state test conditions. The RSD_l estimates were determined for a series of cab leakages from 0.25% to 12%, using instrument RSDs of 0.05 and 0.10 and filter efficiencies of 90%, 95%, and 99%. A mean outside cab concentration of 400 ppm was used for this analysis because it represents an approximate outside atmospheric concentration within which a cab would be tested. Cab penetration leakage Equation A-12 was applied to determine the expected mean cab concentrations for the series of cab leakages at the three filter efficiencies. The expected mean downstream filter concentrations were determined by using the three filter efficiencies and the outside concentration. Mean concentrations inside the cab, outside the cab, and downstream of the filter were assumed to randomly vary with instrument precisions (RSDs) of 0.05 and 0.10. Their standard deviations were estimated by multiplying their expected means by these RSDs.

Figures 11 and 12 show the results of the cab leakage and RSD_l analysis for instrument RSDs of 0.05 and 0.10, respectively, at filter efficiencies of 90%, 95%, and 99%. The individual points shown in Figure 11 are the average leakages and RSD_ls of the instrument test replicates measured on the laboratory test stand (Appendix C) and on the John Deere 7820 tractor cab (Table 7) for the refined leakage testing methodology. These experimental data points are included in Figure 11 because they were measured with outside CO_2 levels near steady-state conditions with outside concentration RSDs near or below 0.05.

Both Figures 11 and 12 clearly show that using a more efficient intake filter and a more precise instrument reduces the cab leakage measurement error (RSD_l). The RSD_l estimates for a 2% leak were 0.43, 0.23, and 0.09 for the 90%, 95%, and 99% filter efficiencies, respectively, assuming an instrument RSD of 0.05. When assuming an instrument RSD of 0.10, the RSD_l

estimates for a 2% leak basically doubles to 0.86, 0.46, and 0.19 for the 90%, 95%, and 99% filter efficiencies, respectively. Both figures also show that the RSD_l estimates for the 90% and 95% filters appear to rapidly increase and diverge from the 99% efficient filter RSD_l estimates when leakage is less than 2%. Conversely, the RSD_l estimates for the 90% and 95% filters converge closer to the 99% filter efficiency estimates when leakage is above 2%. This analysis indicates that cab leakage measurement errors can be reduced by using a higher efficiency filter (> 95% efficiency) and the most precise instrument at or near steady-state conditions. Lower efficiency intake filters and more variable outside concentrations would likely require additional cab testing replicates to increase the confidence in the measurements.

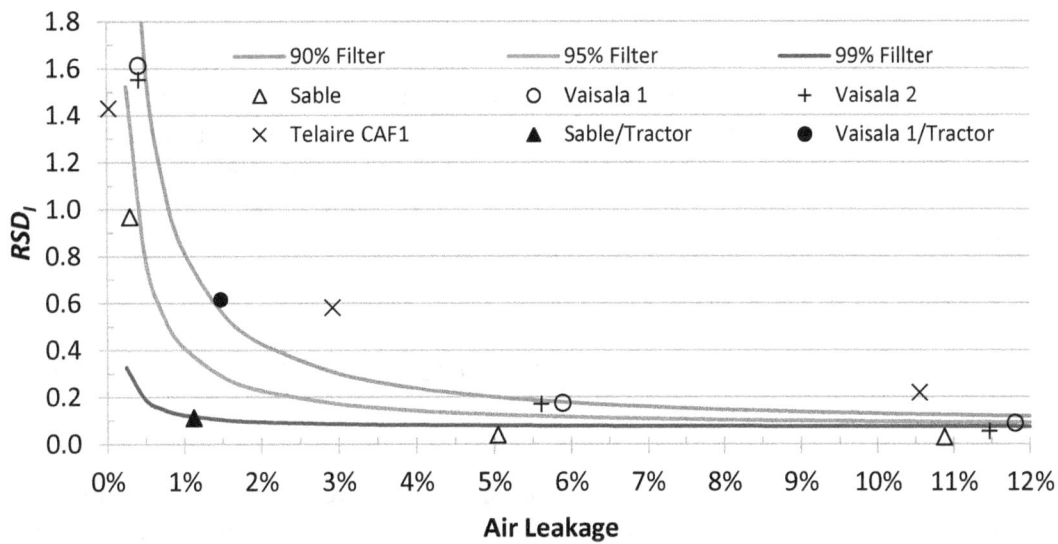

Figure 11. Cab leakage RSD_l estimates with a CO_2 instrument RSD of 0.05.

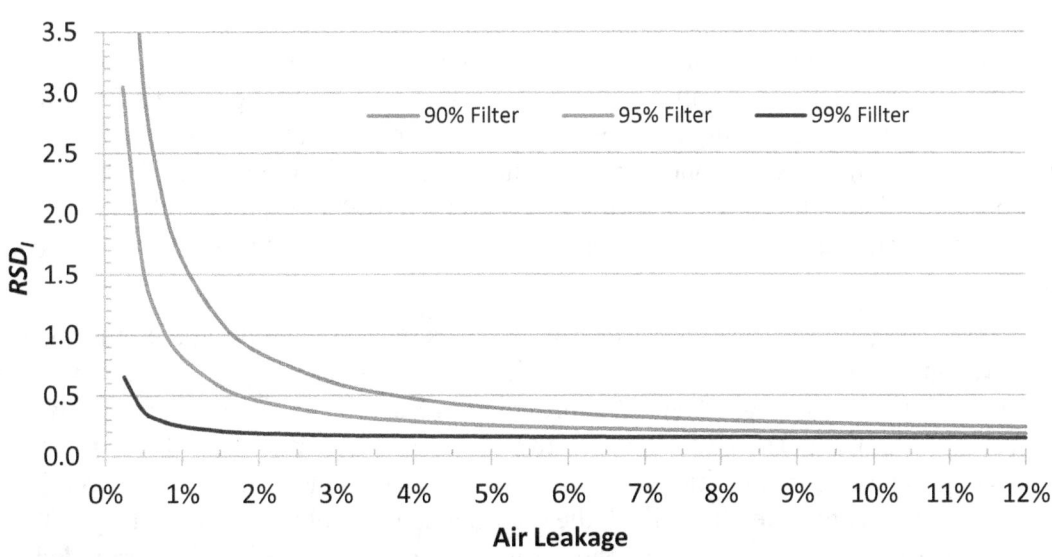

Figure 12. Cab leakage RSD_l estimates with a CO_2 instrument RSD of 0.10.

Figure 11 further illustrates the similarities between the estimated leakage errors, assuming steady-state test conditions and the actual Sable and Vaisala leakage measurement errors conducted near steady-state conditions. The efficiency of the stacked panel filters used on the laboratory test stand was greater than 98% as measured by the Sable (Appendix C). The filter efficiency of the cylindrical filter used during the tractor test was greater than 96% as measured by the Sable instrument (Table 7). As can be seen in Figure 11 the Sable performed closest to the 99% filter efficiency line. The Vaisala overall performed better than the 90% filter efficiency line. The Telaire instrument, in general, did not perform as well as the other instruments. Again this is reflective of the precision of the instruments, with the Sable being the most precise instrument of the three tested and the Vaisala being the second most precise instrument tested.

Finally, the instruments response time can also influence cab leakage measurement error under non-steady-state test conditions. The quicker response time of the Sable (1/2 sec) allows the instrument to more quickly measure concentration changes, while the Vaisala (30 sec) and the Telaire (< 60 sec) are slower to respond to the concentration changes. The slower responding instruments will tend to over or under measure concentration changes that occur more frequently than their response time. Therefore, non-steady-state test conditions can cause additional cab leakage measurement errors for the slower responding instruments because they may never fully complete their measurement response to concentration variations.

Conclusions

Several CO_2 instruments were examined for their suitability in leak testing cab filtration systems. Variations in measurement accuracy and precision were observed during instrument testing. Consistent with manufacturer representations, the Sable was found to be the most accurate and precise instrument tested, within 2 ppm of the reference gases and with a standard deviation less than 1 ppm. It was clear from the testing that all the CO_2 instruments could drift and would need daily checks or calibration to ensure accuracy between the multiple instruments used for cab testing.

A single-instrument, multiple-location cab sampling methodology was devised and examined to eliminate multiple-instrument sampling biases and frequent calibrations during testing. Initial laboratory testing of this sampling methodology showed notable cab test stand leakage discrepancy with changing CO_2 concentrations outside the cab during testing. A more refined test methodology, that reduced the sampling time and increased the sampling frequency at each cab location reduced leakage variations measured under more stable test concentrations outside the cab.

The Sable, being an active sampling instrument, was ideal for sampling the multiple cab filtration system locations and was shown to be the most accurate and precise instrument for the refined cab testing methodology. The Sable provided average leakage measurements within 0.7% of the air quantity measured leakages with 95% confidence intervals $\leq 0.7\%$ for the three complete sampling location repetitions per leakage condition. When the leakage conditions for both test series (replicates) were averaged, the confidence interval was reduced to $\leq 0.3\%$. The Vaisala instruments had to be pump-aspirated from the sampling locations and provided average leakage measurements usually within 1.0% of the air leakage measurements with confidence intervals $\leq 3.3\%$. When both test series under the same leakage conditions (replicates) were averaged together, the Vaisala instruments' confidence intervals decreased to $\leq 1.1\%$. The Telaire instruments also had to be pump-aspirated and provided average leakage measurements greater than 1.0% of the air quantity leakage measurements for about half of the tests conducted with confidence intervals up to 6.1%. Averaging both tests series data (replicates) reduced the Telaire's confidence levels to $\leq 2.4\%$. When compared to the other instruments tested, data indicated that the Telaire was the least suitable for the refined test methodology proposed in this report due to lower accuracy, slower response times, and passive mode of operation.

Further leakage testing with the refined sampling methodology was demonstrated on a John Deere 7820 tractor cab with a special filter cartridge containing sampling ports to the downstream side of the filter. The Sable and Vaisala measured the tractor cab leakage at $1.1 \pm 0.3\%$ and $1.5 \pm 2.3\%$, respectively, during three sampling rotations through the cab filtration system locations. The tractor cab leakage measurement errors for these particular instruments had relatively similar 95% confidence intervals as compared to the cab test stand.

Both the CAF cab test stand experiments and tractor cab results demonstrate that the refined multiple-location cab sampling methodology can be a viable cab leakage test method that does not require frequent CO_2 instrument calibrations during testing. Measuring cab filtration system leakage in the field from airflow quantities is impractical because the filtration system leakage area(s) are usually unknown. Testing cabs for leakage with the CO_2 sampling methodology is best conducted in well-ventilated locations or outside where the CO_2 concentrations are reasonably steady for approximating the steady-state condition assumptions of the mathematical model. A more precise and responsive instrument provides a greater level of measurement certainty when quantifying cab leakages less than or equal to 2%. When less precise instruments are used, or outside cab test conditions fluctuate, additional sampling repetitions may be needed to reduce measured cab leakage confidence intervals. In order to field test cab filtration systems with this sampling methodology, a sampling port needs to be installed within the filter or HVAC system to allow for CO_2 sampling immediately downstream of the filter. Because it was shown that a very high efficiency filter (> 99% efficiency) is not necessary to measure cab leakage using the multiple-location sampling methodology, test filters can be reused multiple times for leakage testing at a somewhat reduced CO_2 absorbent media efficiency performance ($\geq 95\%$ efficiency). The new leak test method provides cab manufacturers, cab service personnel, and industrial hygienists with a measurement tool to ensure environmental cab integrity and minimize worker exposure to outside airborne substances.

References

ASAE (American Society of Agricultural Engineers) [1997]. Agricultural cabs–environmental air quality, Part 1: Definitions, test methods, and safety practices, ASAE Standard S525-1.1, St. Joseph, MI.

Bevington PR [1969]. Data reduction and error analysis for the physical sciences. New York: McGraw-Hill Book Company, Inc., pp. 56–64.

Cecala AB, Organiscak JA, Heitbrink WA, Zimmer JA, Fisher T, Gresh RE, Ashley JD [2003]. Reducing enclosed cab drill operator's respirable dust exposure at a surface coal operation using a retrofitted filtration and pressurization system. In: Yernberg WR, ed. Transactions of the Society for Mining, Metallurgy, and Exploration, Inc. Vol. 314. Littleton, CO: Society for Mining, Metallurgy, and Exploration, Inc., pp. 31–36.

Cecala AB, Organiscak JA, Zimmer JA, Heitbrink WA, Moyer ES, Schmitz M, Ahrenholtz E, Coppock CC, Andrews EH [2005]. Reducing enclosed cab drill operator's respirable dust exposure with effective filtration and pressurization techniques. JOEH *2*:54–63.

Hartman HL [1961]. Mine ventilation and air conditioning. New York: John Wiley & Sons, pp. 37–38.

Heitbrink WA, Hall RM, Reed LD, Gibbons D [1998]. Review of ambient aerosol test procedures in ASAE standard S525. J. Agric. Saf. Health *4*(4):255–266.

Heitbrink WA, Moyer ES, Jensen PA, Watkins DS, Martin Jr. SB [2003]. Environmental agricultural tractor cab filter efficiency and field evaluation. AIHA Journal *64*:394–400.

Heitbrink WA, Collingwood S [2005]. Aerosol generation by blower motors as a bias in assessing aerosol penetration into cabin filtration systems. JOEH *2*(1):45–53.

Knudsen JG, Katz DL [1958]. Fluid dynamics and heat transfer. New York: McGraw-Hill Book Company, Inc., pp. 148–149.

Moyer ES, Heitbrink WA, Jensen PA [2005]. Test for the integrity of environmental tractor cab filtration systems. JOEH *2*:516–523.

NIOSH [1995]. Guidelines for air sampling and analytical method development and evaluation. By Kennedy ER, Fischbach TJ, Song R, Eller PM, Shulman SA. U.S. Department of Health and Human Services, Centers for Disease Control and Prevention, National Institute for Occupational Safety and Health, DHHS (NIOSH) Publication No. 95-117.

NIOSH [2008]. Key design factors of enclosed cab dust filtration systems. By Organiscak JA, Cecala AB. Pittsburgh, PA: U.S. Department of Health and Human Services, Centers for Disease Control and Prevention, National Institute for Occupational Safety and Health, Report of Investigations No. 9677, DHHS (NIOSH) Publication No. 2009–103.

Organiscak JA, Cecala AB, Thimons ED, Heitbrink WA, Schmitz M, Ahrenholtz E [2003]. NIOSH/Industry collaborative efforts show improved mining equipment cab dust protection. In: Yernberg WR, ed. Transactions of Society for Mining, Metallurgy, and Exploration, Inc. Vol. 314. Littleton, CO: Society for Mining, Metallurgy, and Exploration, Inc., pp. 145–152.

Organiscak JA, Schmitz M [2006]. A new concept for leak testing environmental enclosure filtration systems. J. ASTM Int. *3*(10):11 pp.

Organiscak JA, Schmitz M [2009]. Method for leak testing an environmental enclosure. Australian Patent No. 2003286587.

Organiscak JA, Schmitz M [2010]. Method for leak testing an environmental enclosure. United States Patent No. 7,727,765. Washington, DC: U.S. Patent and Trademark Office.

Shanks ME, Gambill R [1973]. Calculus: analytical geometry/elementary functions. New York: Holt, Rinehart, and Winston, Inc., pp. 352–360.

Appendix A. Leakage Model for Cab Filtration System

Development of the cab filtration system leakage model is based on a time-dependent mass balance model of airborne substances in a control volume. Equation A-1 below is a differential equation describing the mass balance of an airborne substance for a cab filtration system control volume as shown in Figure A-1. This is a reformulation of the basic equation for general dilution ventilation [Hartman 1961]. The left-hand part of the equation describes the mass of a particular airborne substance in the control volume. The positive terms in the right-hand part of the equation describe the addition of the mass substance into the control volume from filter penetration and intake air leakage. The negative term describes the removal of the mass substance from the control volume by intake air dilution.

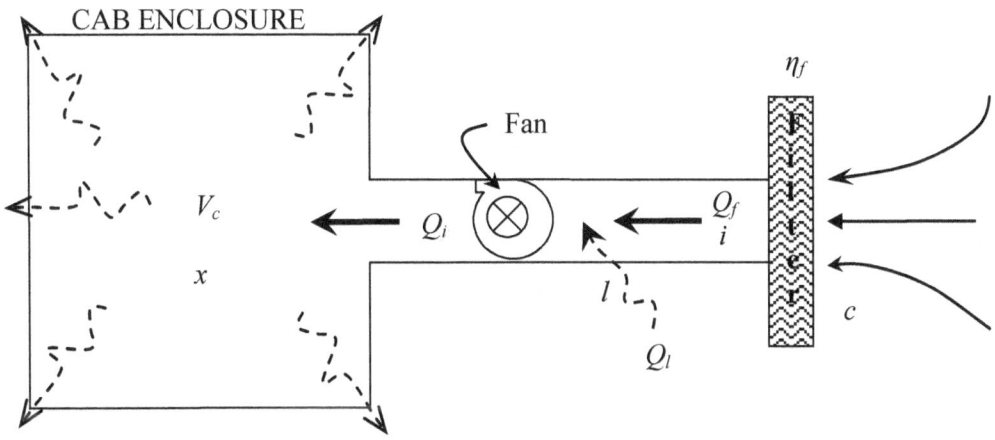

Figure A-1. Basic enclosed cab filtration system.

Mathematical model: $V_c\, dx = c\, Q_l\, dt + c\, Q_f (1 - \eta_f) dt - x\, Q_i\, dt$ (A-1)

Model assumptions:

1. Outside airborne substance concentration is constant.
2. Intake filter removes airborne substance.
3. Airborne substance leakage into the filtration system is proportional to the air quantity leakage around the intake filter.
4. Inside cab volume static pressure is greater than the outside static and wind velocity pressure, so as to keep the outside airborne substance from infiltrating into the cab structure (control volume).

where V_c = cab volume,
  x = inside cab substance concentration,
  c = outside cab substance concentration,
  i = immediate substance concentration after the filter,

η_f = intake filter efficiency, fractional, $(1-i/c)$
Q_f = filtered air quantity, volume per time,
Q_l = air quantity leakage around the intake filter, volume per time,
Q_i = intake air quantity discharged into the cab, volume per time,
l = proportion of cab intake air leakage, or Q_l/Q_i,
t = time,
and Pen = cab penetration, ratio, x/c.

Because: $Q_i = Q_l + Q_f$ and $Q_l = l\,Q_i$; $Q_f = Q_i(1-l)$

Substitute in model:
$$V_c\,dx = c\,l\,Q_i\,dt + c\,Q_i(1-l)(1-\eta_f)dt - x\,Q_i\,dt \qquad \text{(A-2)}$$

Simplify:
$$V_c\,dx = c\,Q_i(1-\eta_f + l\,\eta_f)dt - x\,Q_i\,dt \qquad \text{(A-3)}$$

Rearrange:
$$\int_{x_o}^{x} \frac{dx}{c\,Q_i(1-\eta_f + l\,\eta_f) - x\,Q_i} = \frac{1}{V_c}\int_{t_1}^{t_2} dt \qquad \text{(A-4)}$$

Let: $u = c\,Q_i(1-\eta_f + l\,\eta_f) - x\,Q_i$ and $du = -Q_i\,dx$

Substitute:
$$\frac{1}{-Q_i}\int_{u_o}^{u}\frac{du}{u} = \frac{1}{V_c}\int_{t_1}^{t_2} dt \qquad \text{(A-5)}$$

Integrate:
$$-\frac{1}{Q_i}(\ln|u| - \ln|u_o|) = \frac{(t_2 - t_1)}{V_c} \qquad \text{(A-6)}$$

Rearrange:
$$\ln\frac{u}{u_o} = \frac{-Q_i\,\Delta t}{V_c} \qquad \text{(A-7)}$$

Substitute for u:
$$\ln\frac{c\,Q_i(1-\eta_f + l\,\eta_f) - x\,Q_i}{c\,Q_i(1-\eta_f + l\,\eta_f) - x_o\,Q_i} = \frac{-Q_i\,\Delta t}{V_c} \qquad \text{(A-8)}$$

Antilog of equation:
$$\frac{c\,Q_i(1-\eta_f + l\,\eta_f) - x\,Q_i}{c\,Q_i(1-\eta_f + l\,\eta_f) - x_o\,Q_i} = e^{\left(\frac{-Q_i\,\Delta t}{V_c}\right)} \qquad \text{(A-9)}$$

The steady-state mathematical solution as $\Delta t \to \infty$; $e^{-\infty} \to 0$

Reduces to:
$$c\,Q_i(1-\eta_f + l\,\eta_f) - x\,Q_i = 0 \qquad \text{(A-10)}$$

Rearrange: $$x = \frac{cQ_i(1 - \eta_f + l\eta_f)}{Q_i} \quad \text{(A-11)}$$

Simplify: $$Pen = \frac{x}{c} = 1 - \eta_f + l\eta_f \quad \text{(A-12)}$$

Solve for l: $$l = \frac{\frac{x}{c} + \eta_f - 1}{\eta_f}. \quad \text{(A-13)}$$

This leakage equation can be expressed in CO_2 concentrations measured inside the cab, after the intake filter, and outside the cab as shown below.

Substitute for η_f: $$l = \frac{\frac{x}{c} + 1 - \frac{i}{c} - 1}{1 - \frac{i}{c}}. \quad \text{(A-14)}$$

Simplify: $$l = \frac{\frac{x}{c} - \frac{i}{c}}{1 - \frac{i}{c}}. \quad \text{(A-15)}$$

Rearrange: $$l = \frac{\frac{x-i}{c}}{\frac{c-i}{c}}. \quad \text{(A-16)}$$

Simplify: $$l = \frac{x-i}{c-i} \quad \text{(A-17)}$$

Where: $x \geq i$, $c > i$, and $c \geq x$

Note: This equation is of indeterminate form (0/0) for the particular case where no intake filter is used and all the concentrations are equal ($x = i = c$). In this particular case l' Hospital's rule can be applied as shown below for this indeterminate form [Shanks and Gambill 1973], yielding a limit of 1 or 100% leakage without a filter.

$$\lim_{i \to c} \frac{f(i)}{g(i)} = \lim_{i \to c} \frac{x-i}{c-i} = \lim_{i \to c} \frac{f'(i)}{g'(i)} = \lim_{i \to c} \frac{-1}{-1} = 1$$

Appendix B. Initial Experimental Leakage Data from Cab Test Stand

(Test Conditions @ 972 to 973 hPa, 69.5 to 75.3°F, 22.8 to 24.3% RH)

Test series	Airflow leakage (%)	CO_2 instrument	Cab concentration (ppm)	Filter concentration (ppm)	Outside concentration (ppm)	CO_2 calculated Air leakage (%)
1	0.0%	Sable	7	8	923	-0.2%
			1	6	766	-0.6%
	0.0%	Vaisala 1	46	46	983	0.0%
			39	44	824	-0.6%
	0.0%	Vaisala 2	-6	-4	1023	-0.2%
			-3	2	824	-0.5%
1	5.5%	Sable	59	14	912	5.0%
			53	9	1005	4.4%
	5.5%	Vaisala 1	92	40	952	5.7%
			93	44	1052	4.9%
	5.5%	Vaisala 2	50	1	900	5.5%
			50	-5	985	5.6%
1	10.8%	Sable	93	10	1081	7.8%
			113	16	901	11.0%
	10.8%	Vaisala 1	139	52	1108	8.2%
			157	66	947	10.3%
	10.8%	Vaisala 2	112	17	1018	9.5%
			82	13	927	7.5%
2	0.1%	Sable	2	1	477	0.2%
			0	3	511	-0.6%
	0.1%	Telaire CAF1	2	1	490	0.2%
			4	3	528	0.2%
	0.1%	Telaire CAF2	9	18	525	-1.9%
			14	23	549	-1.5%
	0.1%	Vaisala 2	-17	-24	463	1.5%
			-24	-20	496	-0.9%
2	5.5%	Sable	33	16	461	3.9%
			30	24	487	1.3%
	5.5%	Telaire CAF1	38	26	480	2.8%
			46	36	519	2.0%
	5.5%	Telaire CAF2	51	35	539	3.3%
			66	58	550	1.7%
	5.5%	Vaisala 2	20	-5	447	5.5%
			15	9	476	1.3%
2	10.9%	Sable	52	6	530	8.8%
			53	8	486	9.3%
	10.9%	Telaire CAF1	45	7	544	7.1%
			61	12	506	9.1%
	10.9%	Telaire CAF2	75	24	561	9.5%
			66	38	541	5.7%
	10.9%	Vaisala 2	38	-15	516	5.5%
			37	-10	475	1.3%

Appendix C. Refined Experimental Leakage Data from Cab Test Stand

(Test Conditions @ 961 to 964 hPa, 64.4 to 72.9°F, 45.2 to 52.5% RH)

Test series	Airflow leakage (%)	CO_2 instrument	Cab concentration (ppm)	Filter concentration (ppm)	Outside concentration ppm	CO_2 calculated Air leakage (%)
1	0.0%	Sable	4	5	460	-0.2%
			3	2	419	0.3%
			3	2	415	0.3%
	0.0%	Vaisala 1	47	43	498	0.9%
			48	45	474	0.6%
			47	41	468	1.4%
	0.0%	Vaisala 2	-30	-29	428	-0.2%
			-31	-30	393	-0.2%
			-27	-28	393	0.4%
	0.0%	Telaire CAF1	1	1	445	0.0%
			1	1	412	0.0%
			1	1	408	0.0%
1	5.3%	Sable	23	2	412	5.2%
			23	1	425	5.0%
			23	3	411	5.0%
	5.3%	Vaisala 1	61	40	461	5.0%
			73	41	477	7.3%
			66	40	472	6.0%
	5.3%	Vaisala 2	-8	-33	384	6.0%
			-5	-30	398	5.8%
			-4	-30	388	6.2%
	5.3%	Telaire CAF1	4	1	414	0.7%
			14	1	418	3.0%
			9	3	411	1.5%
1	10.9%	Sable	45	1	412	10.8%
			46	1	396	11.4%
			45	1	399	11.1%
	10.9%	Vaisala 1	91	40	459	12.1%
			91	38	444	12.9%
			90	40	449	12.2%
	10.9%	Vaisala 2	19	-30	388	11.8%
			17	-30	378	11.4%
			21	-30	375	12.6%
	10.9%	Telaire CAF1	29	2	413	6.7%
			37	2	400	8.8%
			48	2	398	11.6%
2	0.0%	Sable	7	4	440	0.7%
			5	4	427	0.2%
			5	4	420	0.3%
	0.0%	Vaisala 1	46	48	485	-0.4%
			44	45	481	-0.2%
			43	42	470	0.2%
	0.0%	Vaisala 2	-26	-33	414	1.5%
			-27	-30	396	0.8%
			-29	-30	392	0.2%
	0.0%	Telaire CAF1	1	1	434	0.0%
			2	2	423	0.0%
			2	2	414	0.1%

Test series	Airflow leakage (%)	CO_2 instrument	Cab concentration (ppm)	Filter concentration (ppm)	Outside concentration (ppm)	CO_2 calculated Air leakage (%)
2	5.1%	Sable	23	4	410	4.7%
			24	4	401	5.2%
			24	4	394	5.2%
	5.1%	Vaisala 1	66	37	461	6.9%
			61	42	448	4.7%
			62	39	446	5.5%
	5.1%	Vaisala 2	-11	-28	383	4.1%
			-11	-31	374	4.9%
			-8	-34	367	6.7%
	5.1%	Telaire CAF1	13	2	405	2.7%
			20	2	397	4.6%
			21	2	385	5.1%
2	10.8%	Sable	44	3	398	10.4%
			46	3	397	10.9%
			46	3	398	10.8%
	10.8%	Vaisala 1	83	43	444	10.0%
			90	40	447	12.3%
			90	45	443	11.3%
	10.8%	Vaisala 2	10	-34	373	10.9%
			15	-30	373	11.2%
			10	-35	371	11.1%
	10.8%	Telaire CAF1	45	2	386	11.4%
			49	2	386	12.2%
			50	2	389	12.6%

Appendix D. Propagation of Error Analysis for the Cab Leakage Testing Methodology

Instrument imprecision and unsteady test conditions increases the propagation of measurement error for the single-instrument, multiple-location cab sampling leakage method. In order to examine these measurement errors, the following uncertainty analysis for this testing methodology was conducted below to show the relative standard deviation of cab leakage measurements [Bevington 1969].

Cab Leakage Equation:

$$l = f(x, i, c) = \frac{x-i}{c-i} \qquad (A\text{-}17)$$

The variance for this three variable cab leakage equation is:

$$s_l^2 = s_x^2 \left(\frac{\partial l}{\partial x}\right)^2 + s_i^2 \left(\frac{\partial l}{\partial i}\right)^2 + s_c^2 \left(\frac{\partial l}{\partial c}\right)^2 + 2 s_{xi}^2 \left(\frac{\partial l}{\partial x}\right)\left(\frac{\partial l}{\partial i}\right) + 2 s_{xc}^2 \left(\frac{\partial l}{\partial x}\right)\left(\frac{\partial l}{\partial c}\right)$$
$$+ 2 s_{ci}^2 \left(\frac{\partial l}{\partial c}\right)\left(\frac{\partial l}{\partial i}\right) \qquad (D\text{-}1)$$

where l = mean of proportional leakage,
s_l = standard deviation of proportional leakage,
x = mean cab concentration,
s_x = standard deviation of cab concentration,
i = mean downstream filter concentration,
s_i = standard deviation of downstream filter concentration,
c = mean outside cab concentration,
s_c = standard deviation of outside cab concentration,
s_{xi}^2 = covariance between x and i,
s_{xc}^2 = covariance between x and c,
s_{ci}^2 = covariance between c and i.
RSD_l = relative standard deviation of leakage estimate, s_l/l,

and
$$\left(\frac{\partial l}{\partial x}\right) = \frac{(c-i)(1)-(x-i)(0)}{(c-i)^2} = \frac{c-i-0}{(c-i)^2} = \frac{1}{c-i}$$

$$\left(\frac{\partial l}{\partial i}\right) = \frac{(c-i)(-1)-(x-i)(-1)}{(c-i)^2} = \frac{i-c+x-i}{(c-i)^2} = \frac{-(c-x)}{(c-i)^2}$$

$$\left(\frac{\partial l}{\partial c}\right) = \frac{(c-i)(0)-(x-i)(1)}{(c-i)^2} = \frac{-(x-i)}{(c-i)^2}$$

Substitute:

$$s_l^2 = s_x^2 \left(\frac{1}{(c-i)}\right)^2 + s_i^2 \left(\frac{-(c-x)}{(c-i)^2}\right)^2 + s_c^2 \left(\frac{-(x-i)}{(c-i)^2}\right)^2 + 2s_{xi}^2 \left(\frac{1}{(c-i)}\right)\left(\frac{-(c-x)}{(c-i)^2}\right)$$
$$+ 2s_{xc}^2 \left(\frac{1}{(c-i)}\right)\left(\frac{-(x-i)}{(c-i)^2}\right) + 2s_{ci}^2 \left(\frac{-(x-i)}{(c-i)^2}\right)\left(\frac{-(c-x)}{(c-i)^2}\right)$$

(D-2)

Simplify:

$$s_l^2 = s_x^2 \frac{1}{(c-i)^2} + s_i^2 \frac{(c-x)^2}{(c-i)^4} + s_c^2 \frac{(x-i)^2}{(c-i)^4} - 2s_{xi}^2 \frac{(c-x)}{(c-i)^3} - 2s_{xc}^2 \frac{(x-i)}{(c-i)^3}$$
$$+ 2s_{ci}^2 \frac{(x-i)(c-x)}{(c-i)^4}$$

(D-3)

Re-expressed symmetrically:

$$RSD_l^2 = \frac{s_l^2}{l^2} = s_x^2 \frac{1}{(c-i)^2}\left(\frac{c-i}{x-i}\right)^2 + s_i^2 \frac{(c-x)^2}{(c-i)^4}\left(\frac{c-i}{x-i}\right)^2 + s_c^2 \frac{(x-i)^2}{(c-i)^4}\left(\frac{c-i}{x-i}\right)^2$$
$$- 2s_{xi}^2 \left(\frac{c-x}{(c-i)^3}\right)\left(\frac{c-i}{x-i}\right)^2 - 2s_{xc}^2 \left(\frac{x-i}{(c-i)^3}\right)\left(\frac{c-i}{x-i}\right)^2$$
$$+ 2s_{ci}^2 \left(\frac{(x-i)(c-x)}{(c-i)^4}\right)\left(\frac{c-i}{x-i}\right)^2$$

(D-4)

Simplify:

$$RSD_l^2 = \frac{s_l^2}{l^2} = s_x^2 \frac{1}{(x-i)^2} + s_i^2 \frac{(c-x)^2}{(c-i)^2(x-i)^2} + s_c^2 \frac{1}{(c-i)^2} - 2s_{xi}^2 \frac{(c-x)}{(c-i)(x-i)^2}$$
$$- 2s_{xc}^2 \frac{1}{(c-i)(x-i)} + 2s_{ci}^2 \frac{(c-x)}{(c-i)^2(x-i)}$$

(D-5)

Take the square root to solve for the relative standard deviation of the leakage:

$$RSD_l = \frac{s_l}{l} = \sqrt{\begin{array}{c} s_x^2 \frac{1}{(x-i)^2} + s_i^2 \frac{(c-x)^2}{(c-i)^2(x-i)^2} + s_c^2 \frac{1}{(c-i)^2} - 2s_{xi}^2 \frac{(c-x)}{(c-i)(x-i)^2} \\ -2s_{xc}^2 \frac{1}{(c-i)(x-i)^2} + 2s_{ci}^2 \frac{(c-x)}{(c-i)^2(x-i)} \end{array}} \qquad \text{(D-6)}$$

www.ingramcontent.com/pod-product-compliance
Lightning Source LLC
Chambersburg PA
CBHW081804170526
45167CB00008B/3324